THE ART OF GEOLOGY

Edited by

Eldridge M. Moores
Department of Geology
University of California
Davis, California 95616

F. Michael Wahl
Geological Society of America
3300 Penrose Place, P.O. Box 9140
Boulder, Colorado 80301

Special Paper 225
Published by
The Geological Society of America, Inc.
3300 Penrose Place, P.O. Box 9140
Boulder, Colorado 80301

Published by The Geological Society of America, Inc.,
3300 Penrose Place, P.O. Box 9140, Boulder, Colorado 80301

Printed in U.S.A.

Library of Congress Cataloging-in-Publication Data

The Art of Geology.

    (Special paper ; 225)
      1. Geology — Pictorial works.   2. Landforms —
      Pictorial works.   I. Moores, Eldridge M., 1938-
      II. Wahl, F. Michael, 1931-   III. Geology
      (Boulder).   IV. Geological Society of America.   V. Series:
      Special papers (Geological Society of America) ; 225.
QE33.A77   1988   551.4'022'2   88-24533
ISBN 0-8137-2225-X

10 9 8 7 6 5 4 3 2

Contents

Acknowledgments

his volume would not have been possible without the contributions and assistance of many people. First and foremost the editors thank all the earth scientists who submitted photographs for consideration. The uniformly high quality of the submissions made the final selection process both difficult and rewarding for us.

E-An Zen of the United States Geological Survey, Washington, D.C., and a member of the GSA Council, originally suggested the idea for this volume, as a collection of color covers of the GSA journal *Geology*. Faith Rogers, Managing Editor of *Geology*, gave valuable assistance in selection of photographs, editing of text, and organization of the volume. James R. Clark, Production Manager of GSA, helped organize the project, maintained contact with the contributors, and helped select the photographs. Susan Upson, Assistant Managing Editor of *Geology*, helped in selection and text editing. Judith E. Moores, Faith E. Riker, Richard Cowen, and Jeanne Bernauer read one or more versions of the Introduction and made many suggestions for its improvement.

To all these people we extend our heartfelt thanks.

Contributors

David J. Anastasio

Robert S. Anderson

Suzanne Prestrud Anderson

Robert A. Baird

Steve Barnett

Chester B. Beaty

K. T. Biddle

Dennis K. Bird

Ronald C. Blakey

Peter K. Blomquist

Manuel G. Bonilla

Carlton E. Brett

R. L. Brovey

Peter L. Butterworth

Zbigniew Bzdak

D. W. Caldwell

Donald B. Campbell

Renee D. Carver

Rodney Catanach

Dana Q. Coffield

Sterling S. Cook

L. S. Crumpler

H. Allen Curran

Jon P. Davidson

George H. Denton
William J. Devlin
Robin P. Diedrich
David P. Dockstader
Hellmut H. Doelling
Russell F. Dubiel
L. H. Fairchild
Ivan P. Gill
J. D. Griggs
Giovanni Guglielmo, Jr.
Bernard Hallet
Russell S. Harmon
James W. Head
Roland Hellmann
A. Hoppe
Roderick A. Hutchinson
James Irwin
Glen A. Izett
Thomas M. Jacob
Mark J. Johnsson
Wallace D. Kleck
Bart Joseph Kowallis
Mary M. Lahren
Julie E. Laity
David A. Lawler
Irene S. Leung
Peter W. Lipman
Baerbel K. Lucchitta
Richard P. Major
Craig E. Manning
Gail L. Marquardt
R. Mark Maslyn
Niall J. Mateer
James McCalpin
Alfred S. McEwen
David W. McGarvie
Nancy J. McMillan
Peter Michael
Martin G. Mille
Scott H. Miller
Thomas P. Miller
Siegfried Muessig
Kenneth R. Neuhauser
Jeffrey B. Plescia

Wayne Ranney
Kenneth A. Rasmussen
John W. Robinson
Nicholas M. Rose
Mark Savoca
David A. Sawyer
Winsried Schmidt
Richard A. Schweickert
Stephen Self
G. Shanmugam
John S. Shelton
Douglas Smith
Laurence A. Soderblom
Lawrence R. Solkoski
Steven G. Spear
Bruce Taterka
Lyn Topinka
George Ulrich
Martin J. Van Kranendonk
Jennifer Whipple
M. Woodbridge Williams
Grant C. Willis
Monte D. Wilson
Gerhard Wörner
David K. Yamaguchi
Donald H. Zenger

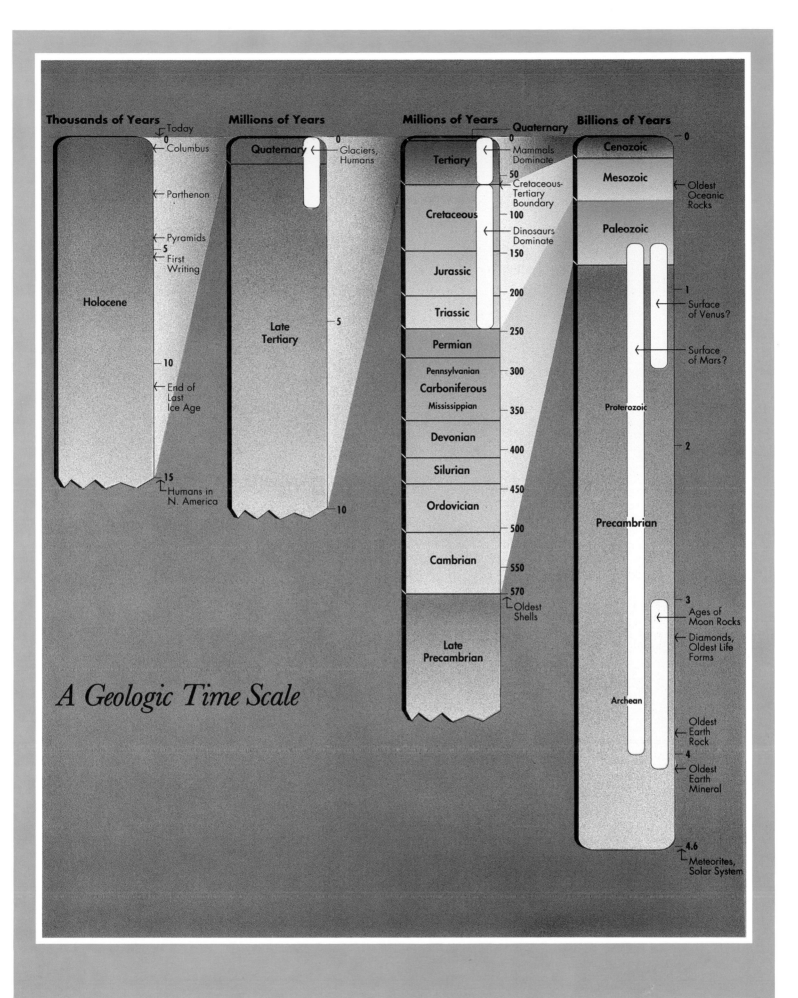

Thousands of Years

Today
Columbus
Parthenon
Pyramids
First Writing

Holocene

End of Last Ice Age

Humans in N. America

Millions of Years

Quaternary — Glaciers, Humans

Late Tertiary

Millions of Years

Quaternary
Tertiary — Mammals Dominate
Cretaceous-Tertiary Boundary
Cretaceous
Dinosaurs Dominate
Jurassic
Triassic
Permian
Pennsylvanian
Carboniferous
Mississippian
Devonian
Silurian
Ordovician
Cambrian
Oldest Shells

Late Precambrian

Billions of Years

Cenozoic
Mesozoic — Oldest Oceanic Rocks
Paleozoic
Surface of Venus?
Surface of Mars?
Proterozoic
Precambrian
Ages of Moon Rocks
Diamonds, Oldest Life Forms
Archean
Oldest Earth Rock
Oldest Earth Mineral
Meteorites, Solar System

A Geologic Time Scale

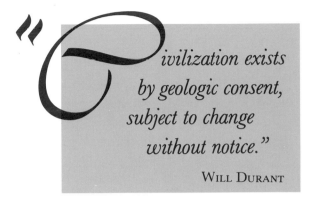

"ivilization exists by geologic consent, subject to change without notice."

WILL DURANT

This volume has two purposes: to celebrate the 100th anniversary of the Geological Society of America and to convey the visual beauty of Earth and its neighbors as seen from a geologic perspective.

The Geological Society of America is a professional society of some 16,000 earth scientists dedicated to promotion of the science of geology in North America. It issues scholarly publications, holds scientific meetings, and provides support for geologic research. Its two monthly journals include the *Geological Society of America Bulletin,* a traditional journal published since 1888, and *Geology,* a journal of short articles on new, fast-breaking developments, published since 1973. The color covers of *Geology* inspired this volume and form its core.

The Society was founded in Ithaca, New York, in December, 1888. At the time a "lame duck" President, Grover Cleveland, was in the White House, Sir John Macdonald was Prime Minister of the Dominion of Canada, and Porfirio Diaz was the President of Mexico. The United States of America consisted of some 63 million people in 38 states and several territories. Spain ruled the Philippines, Marianas, Cuba, and Puerto Rico, and King Kalakaua reigned over the independent country of Hawaii.

In 1888, the first electric lights had only recently been installed in London and New York. Despite the completion of the first U.S. (1869) and Canadian (1885) transcontinental railroads, crossing the North American continent was still a two-week ordeal. Henry Ford was 26, Guglielmo Marconi 14, Albert Einstein 9, the Wright brothers 21 and 17, and Alfred Wegener 6. Their subsequent contributions to automobiles, radio, physics, airplanes, and continental drift lay in the future.

The world reeled from the implications of Charles Darwin's work, although the great man himself had died six years earlier. On the basis of the work of Darwin and other nineteenth century naturalists, geologists had erected a geologic time scale of the relative ages of fossil-bearing strata and had established time periods (Paleozoic, Mesozoic, Cenozoic, and their subdivisions, as seen in the accompanying time line). Geologists had no real idea, however, of the actual ages of these time periods. The great American reconnaissance surveys of the West had ended, but no systematic geologic knowledge of the continent existed.

Five years before GSA was founded, the great volcanic eruption of Krakatau had killed tens of thousands of people in Indonesia and affected the weather around the world for months. A strong earthquake had badly damaged Charleston, South Carolina, only two years before, but the San Francisco quake lay nearly two decades in the future. Little was known about the geologic causes of these natural disasters.

Physicists and geologists were locked in a controversy about the age of Earth. Physicists, led by Lord Kelvin, argued that as little as 10 million years could have elapsed since Earth had first formed, because otherwise it would have cooled off to the point where no volcanic eruptions could occur. Throughout the 19th century geologists had argued that the rocks exposed on Earth's surface were formed by observable processes such as rivers, waves, and volcanoes. The rate of deposition of, for example, sand on a beach, extrapolated to the known total thickness of sediment, implied that Earth was hundreds of millions or a few billion years old.

In the past century, the world has experienced two world wars and tremendous technological changes in communication, transportation, and extraterrestrial exploration. Concurrently, geology has experienced revolutionary change occasioned by radioactive dating of rocks and meteorites; exploration of the oceans, resulting in a new view of Earth—the theory of plate tectonics; geologic observations of Venus, Mars, the Moon, Earth itself, and other planets from spacecraft; and application of the principles of evolution and biochemistry to the fossil record.

The discovery of radioactivity in 1896 made it clear that some of the heat in Earth's interior could result from radioactive processes that release energy, thus invalidating Kelvin's arguments. Estimates of the age of Earth and the Solar System increased from a few tens of millions of years in 1888 to 1.5 billion years by World War I, 3 billion years by World War II, and finally 4.6 billion years by 1960. Surprisingly, the geologic time scale devised in the 19th century occupies only a small part of Earth's historical record. The oldest shelly fossils are in rocks about 570 *million* years old; thus they record only about the last one-eighth of Earth history. The oldest life forms—single celled fossils—are about 3.3 billion years old.

The most exciting developments in geology in the past 25 years have centered around the theory of plate tectonics. We now recognize that the outer part of Earth is composed of a series of segments, or plates, that are in motion with respect to each other. Plates move apart at ridges in the middle of oceans, they slide past each other along fault zones like the San Andreas fault of California, and they converge in regions such as the Pacific Ring of Fire where one plate descends beneath another. Nearly all earthquakes and volcanoes are located at plate boundaries. Continents drift about as plates move, diverging in some places and coming together in others. As they move apart, features such as the Great Basin of the western United States or the Red Sea form, and, later, oceans such as the Atlantic develop. Where plates come together, mountain ranges such as the Alps or the Himalaya are created. For example, the Appalachians developed some 300 million years ago as Africa approached and collided with North America.

The oldest rocks in the deep oceans are only about 180 million years old; older ones have disappeared by subduction of one plate beneath another. The oldest continental rocks are about 3.8 billion years old; the oldest minerals are about 4.1 billion years old. Moon rocks, on the other hand, range from about 3.0 to 4.1 billion years. Surface features on Mars are thought to be approximately 0.1 billion to 3.9 billion years old; features on Venus are about 500 to 1000 million years old.

Geology involves the application of physics, biology, and chemistry to the study of the past and present Earth and other planetary bodies. This study adds the fourth dimension of time to the three dimensions of space. Geologists study features ranging in size from submicroscopic up to entire planets. The smallest items illustrated in this book are microscopic views perhaps 0.1 millimetre* across. The largest view is part of the Martian surface several thousand kilometres across. One thousand kilometres is ten billion times as large as 0.1 millimetre. (One can get some idea of how large one billion is by keeping in mind that one billion millimetres is 1000 kilometres, roughly the distance from San Francisco to Salt Lake City, or from New York to Chicago. Ten billion inches is approximately the distance from Earth to the Moon.)

The time duration of geologic processes shows a similar variation. Events such as earthquakes or volcanic eruptions can be over in minutes, whereas the opening of the Atlantic Ocean has lasted 180 million years, and the continents were mostly formed by 2.5

*In keeping with the standard practice of *Geology* magazine, the international spellings of weights and measures have been used in this volume (*e.g.,* metre, not meter).

billion years ago. (One million seconds is 11.5 days; one billion seconds is about 30 years!)

Geology is an extension of history and archeology to a much longer time scale and a less well preserved record. It is a science that has the whole Earth and other planetary bodies as its laboratory. Geologic features know no political frontiers. Proper understanding of the fragmentary and continuously changing record requires a global overview, because some parts of the record may be better preserved in one part of the world than another. Thus, knowledge of rocks in east Asia or the eastern Mediterranean region may aid in understanding rocks in, for example, British Columbia, New York, or Yucatan. The best geologists develop an intuition about how rocks behave and can often make sound judgments based upon very limited data.

Thus, geologists appreciate Earth and her neighbors from three different perspectives—scientific, human, and esthetic. The scientific perspective involves an understanding of the planets and how and why they change and evolve. Do other planets, for example, show evidence of the same processes of erosion and change or for plate tectonics as we see on Earth? The human perspective involves the geology responsible for hazards or the concentration of mineral resources. For example, why are earthquakes and volcanic eruptions where they are? Why are petroleum resources so abundant in the Middle East or metal deposits in southern Africa? How have these geologic facts affected the developments of civilization? The esthetic level involves the natural beauty of the subjects we study, such as the mountains, the lakes, the river valleys, the shore, and even crystals under a microscope.

The photographs in this book have been chosen both for their visual appeal and for the stories they tell. They give some idea of the diversity of geology and of the enjoyment geologists experience in their work.

E. M. Moores

Slot Canyons, Colorado Plateau, U.S.A.

cattered through hidden localities on the Colorado Plateau in Utah and Arizona are erosional features known as "slot canyons." Many are carved in Jurassic and perhaps Triassic (early Mesozoic 144 to 245 million years old) sandstone. Wind deposited the sand from which these rocks formed. The massive cliff-forming sandstone is commonly light-brown to white, or buff to gray; locally, it can have pink, red, and yellow banding. Well-developed, large-scale cross-stratification indicates that the sandstone has an eolian (wind-blown) origin.

The canyons have resulted from downcutting of streams in response to rapid uplift of the land over the past 10 million years. Geologists have determined that much of western North America has risen rapidly, but they don't know why.

Photo by
Jennifer Whipple

Photo by
Jennifer Whipple

Slot Canyons, Colorado Plateau, U.S.A.

Photos by
Jennifer Whipple

Svartifoss Waterfall, Iceland

At the Svartifoss waterfall in the Skaftafell National Park in southeast Iceland, the Boejargil River cascades over a magnificent exposure of columnar basalt. The columns are as much as 15 metres long and 1 metre in diameter. Columnar basalts are formed by cooling and cracking of lava flows from volcanic eruptions. Iceland, about the same size as West Germany or Pennsylvania, is an island formed by volcanic activity on the ocean floor.

Photos by
Roland Hellmann

Andes Mountains, Peru

The Andes Mountains owe their great height to the relative movement between two tectonic plates, the moving, rigid segments of Earth. The Nazca plate, the part of Earth's crust under the Pacific Ocean, is descending under the South American plate at a rate of a few centimetres per year. This process, called subduction, has resulted in thickening of the western edge of South America, and in the region of Machu Picchu has produced peaks 6000 metres high. The Urubamba River, on its way to the Atlantic Ocean via the Ucayali and Amazon rivers, has cut a deep canyon through this high country.

A sharp bend in the Urubamba River forms a steep-sided promontory that is accessible from only one direction, and the site has rock suitable for masonry, as well as a source of water. At this secluded location, the Incas established the settlement of Machu Picchu.

The masonry structures of Machu Picchu were constructed of stone quarried at the site. To make building stones, the Incas made openings into which they drove wooden wedges. Wetting of the tightly driven wedges caused them to swell and split the rock.

Photos by
Manuel G. Bonilla

Tambora Volcano, Indonesia

 he April 1815 eruption of Tambora Volcano, Indonesia, was one of the largest since the last Ice Age (10,000 years ago). Tambora is part of the Pacific "Ring of Fire"—volcanoes located over subducting plates.

The satellite image of the volcano shows 1815 deposits in blue and magenta. The circular caldera is about the same size as Crater Lake, Oregon.

The space shuttle photograph of the lesser Sunda Islands, Indonesia, shows Tambora on Sumbawa at bottom right. Island volcano with eruptive plume is Sangeang Api.

The volume of the 1815 eruption was 50 cubic kilometres, or 50 times that of the 1980 Mount St. Helens eruption.

Atmospheric dust and aerosols (tiny droplets of sulfuric acid) from this eruption blocked sunlight in the Northern Hemisphere for many months, and caused the "year without a summer" in 1816. The dust also produced spectacularly colored sunrises and sunsets, perhaps as illustrated in the painting of an English landscape by the British artist J.M.W. Turner.

Grain and vegetable crops failed in New England and northern Europe. Inhabitants of entire towns in Massachusetts and Maine left their homes and migrated to the midwestern United States.

Submitted by
Stephen Self
Photo courtesy NASA

Photo by D.W. Caldwell

Photo by D.W. Caldwell

Photo by D.W. Caldwell

Tufa Pinnacles, California, U.S.A.

"Mysterious" and "weird" describe the tufa pinnacles at the south end of Mono Lake, California. Mono Lake, on the east side of the Sierra Nevada, is a shallow salty lake with no outlet. The city of Los Angeles has diverted water from the streams feeding Mono Lake; this has caused the lake to shrink and become saltier.

Tufa deposits have formed by the interaction of fresh-water springs percolating up and mingling with the alkaline, carbonate-bearing lake water.

Photo by
Renee D. Carver

Tufa Pinnacles,
California, U.S.A.

Sorted Circles, Spitsbergen

The graceful curves of small gravel ridges pattern vast areas of the ground at Kvadehuks-letta, on the Arctic island of Spitsbergen. These patterns are found in regions where the ground freezes and thaws annually. The largest and best developed sorted circles form in mixtures of fine-grained material and gravel, where water is abundant, and where the ground just below the surface is permanently frozen.

Photos by Suzanne Prestrud Anderson and Bernard Hallett with the cooperation of the Norsk Polarinstitutt.

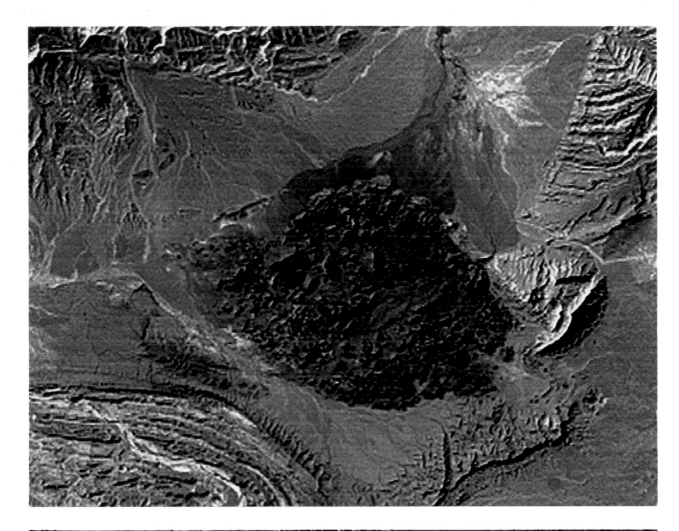

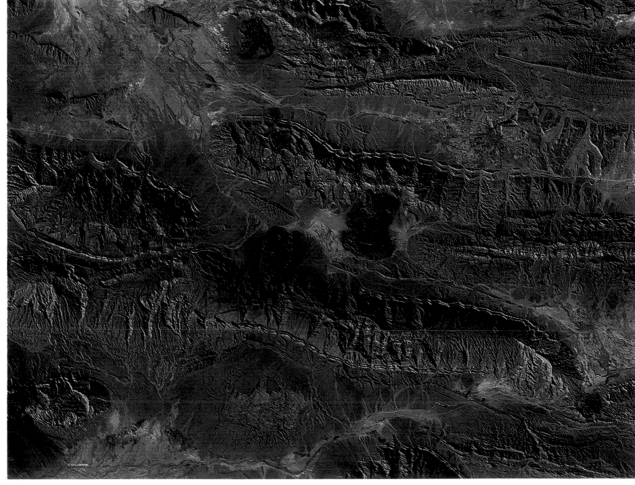

Zagros Mountains, Iran

hese satellite scenes show part of the Zagros fold belt near the Straits of Hormuz, southern Iran. Dark gray areas are salt plugs, formed by Precambrian to early Paleozoic age (550 to 650 million years old) salt that flowed to the surface after piercing through the overlying rocks.

As the salt is extruded, it flows downslope as "glaciers."

The beautiful expression of the Zagros folds and the emergent salt plugs reflect the youth of the folds, formed during the collision of Arabia and Iran, in the past 10 million years, and the arid climate.

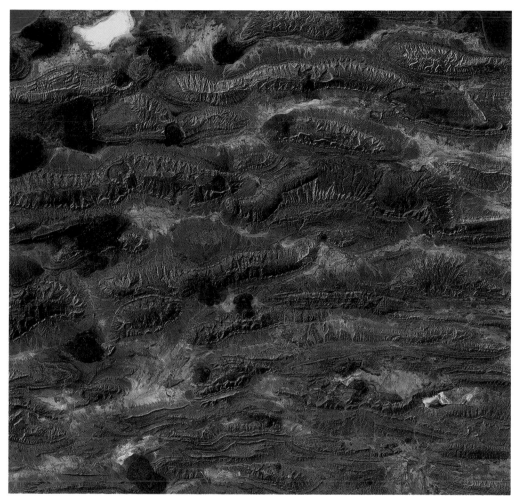

Landsat photos processed at Exxon Production Research Company

Kilauea Volcano, Hawaii, U.S.A.

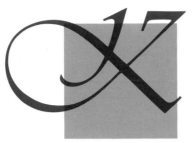

Kilauea volcano is on the southern coast of the island of Hawaii. The islands of Niihau, Kauai, Oahu, Molokai, Lanai, and Maui form a chain northwest of Hawaii, and they represent a "trail" of past eruptions caused by the Pacific plate moving northwestward over a stationary "hot spot" (volcanic center) located in Earth's upper mantle. In November and December 1986, lava poured out of a vent named Puu Oo; streams of molten basalt pushed through the communities of Royal Gardens and Kalapana, destroying nearly everything in their path to the sea. Trees were burned to ashes, leaving holes where they once stood. The heat was so intense that windows of trapped cars and trucks melted over the steering columns.

Photo by George Ulrich courtesy U.S. Geological Survey

Photo by J. D. Griggs, courtesy U.S. Geological Survey

Photo by George Ulrich
courtesy U.S.
Geological Survey

Photo by George Ulrich
courtesy U.S.
Geological Survey

Kilauea Volcano, Hawaii, U.S.A.

Photo by George Ulrich

Photo by Kenneth R. Neuhauser

Photo by
Kenneth R. Neuhauser

Photo by
Kenneth R. Neuhauser

Zumaya Flysch, Spain

This sequence of flysch is exposed near the town of Zumaya, northern Spain. The rock formed from sediment deposited on the ocean floor by bottom currents near the margins of continents. The originally horizontal layers were tilted steeply during mountain-building movements over the past few tens of millions of years.

Photos by
G. Shanmugam

Teton National Park, Wyoming, U.S.A.

The eastern face of the Teton Mountains formed by uplift of the range along a steep fault; it has been sculpted by glaciers during the past million years. The Snake River meanders in a valley eroded into old river gravels. Although the glaciers that carved the Tetons into rugged spires have retreated, they are not entirely gone; small icefields can still be seen on high north-facing benches.

Photos by
James McCalpin

Photos by
David R. Dockstader

Taughannock Falls, New York, U.S.A.

ew York State's highest waterfall, Taughannock Falls, forms where Taughannock Creek tumbles over the edge of the Lockport dolomite, through the soft shales beneath, and into the valley of Cayuga Lake (one of the Finger Lakes). Glaciers carved the deep valley of Cayuga Lake only 10,000–20,000 years ago. The Lockport dolomite, which creates the hard cap of the valley rim, comes from reef deposits that formed in a warm shallow sea during the Silurian period (mid-Paleozoic, about 400 million years ago). Small, vertical fractures or joints in the Lockport dolomite control the course of Taughannock Creek and form the lip of the falls.

Mayon Volcano, Philippines

ayon, Luzon Island, Philippines, is the most active volcano in the Philippine part of the Pacific "Ring of Fire." The first recorded eruption of Mayon was in 1616, and it erupts every ten years or so. Mayon may be the heat source for hot-springs systems nearby. Gold is mined in this area of volcanoes, mangrove swamps, rice paddies, and coconut plantations.

Photos by
Lawrence R. Solkoski

Photos by
Wayne Ranney

Unconformities, Grand Canyon, Arizona, U.S.A.

*G*eologists study the rock record to help unravel the mysteries of Earth history. And yet this record records only a small fraction of all geologic time. Much of Earth history occurred "between" the emplacement of various crystalline rocks or the deposition of different layers of sediment. These gaps in the record, known as *unconformities,* are the result either of a period when sediments are not deposited or of the uplift and erosion of an older deposit before the next one is laid.

The Grand Canyon of the Colorado River in northern Arizona is an excellent place to view these gaps in the rock record. One gap, known in the Grand Canyon as the Great Unconformity, represents 1.25 billion years of Earth history and is represented by the middle Cambrian (early Paleozoic) horizontal strata, about 540 million years old, which overlie the Precambrian crystalline rocks of the inner gorge. During this period, the crystalline rocks were lifted into a lofty mountain range; these mountains were eroded to a flat lowland plain; over 12,000 feet of sedimentary rocks accumulated on this plain, and were then lifted, faulted, and eroded to yet another lowland plain cut mostly into crystalline bedrock. These events, and the mind-boggling span of time they represent, are contained within the thickness of a fingernail in the Grand Canyon. The Grand Canyon is a natural and geologic wonderland, but much of its story is contained in the rocks that are *not* there!

Limestone Caves, U.S.A. and Brazil

Photo by
Mark J. Johnsson

*L*imestone caves start to form when acidic rainwater falls, percolates into the ground, and dissolves the rock. Caverns form along bedding planes and fractures in the rock and are enlarged by underground stream erosion. When the water becomes saturated with calcium carbonate, new deposits form in the caverns—columns, curtains, and crusts.

Photo by
R. Mark Maslyn

Photo by
Giovanni Guglielmo, Jr.

Photo by
Giovanni Guglielmo, Jr.

33

Limestone Caves,
U.S.A. and Brazil

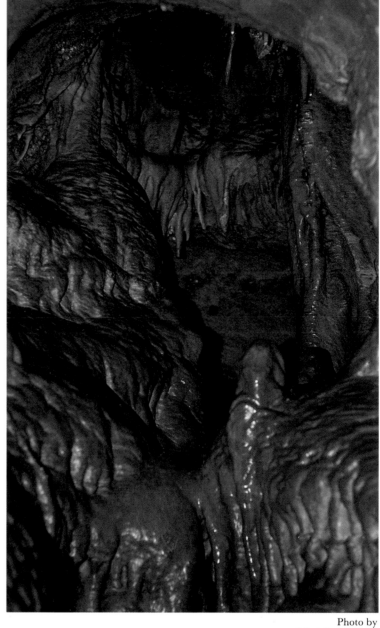

Photo by
Mark J. Johnsson

Photo by
Giovanni Guglielmo, Jr.

Thrust Faults, Spain

*M*ost mountain belts form when continents collide. These collisions cause the continental margins to wrinkle and rise. Thrust faults, which form when rocks on one side of a break move up and over the other side, are the dominant structures in these mountain belts. The External Sierra of the Spanish Pyrenees displays well-preserved examples of thrust faults that formed about 20 million years ago as Spain ceased colliding with France.

Photos by
David J. Anastasio

Recent Glacial Retreat, Alaska, U.S.A.

Since the end of the last major glacial period some 10,000 years ago, most of the mountain glaciers in the world have been retreating, as winter snowfall fails to keep up with summer melting. A hundred years ago, these three glaciers near Portage, southeast of Anchorage, Alaska, were a single glacier, which extended across the lake in the foreground.

Photos by
Steven G. Spear

Cathodoluminescent Bimineralic Ooids, Florida, U.S.A.

oids are sand-sized grains that form when calcium carbonate precipitates around a nucleus in warm, tropical surf zones. These ooids from the Florida continental shelf formed several thousand years ago, during the last glacial period, when sea level was lower.

Viewing a thin, transparent slice of the rock in a polarizing micro-scope reveals the concentric structure of the ooids. These pictures show views of such a slice produced by using three kinds of light: (1) plane-polarized light, which reveals textures between individual minerals; (2) cross-polarized light, which helps identify different minerals by producing interference color patterns; and (3) cathodo-luminescence, in which the ooids are placed in a vacuum and bombarded with electrons; trace amounts of manganese in the mineral crystal cause the ooids to emit an orange glow.

Studying ooids this way helps answer questions about how they formed and how they subsequently were changed chemically.

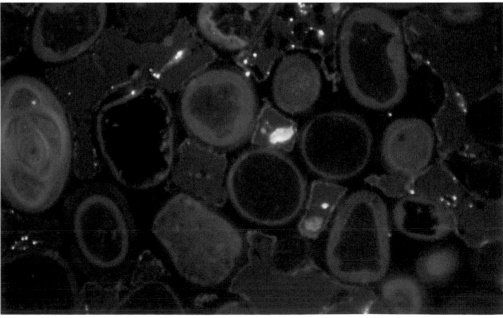

3

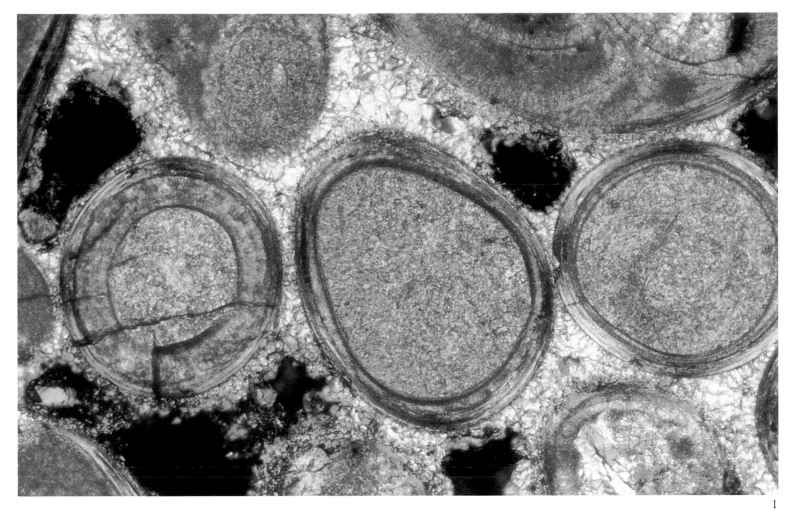

1

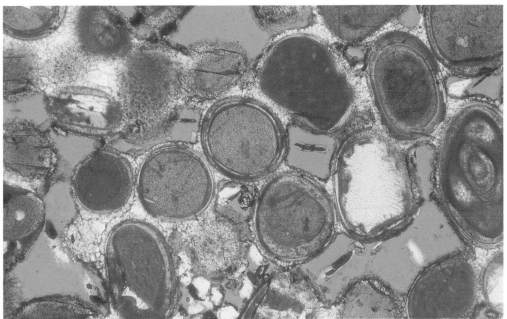

Photos by R. P. Major

2

Photos by Peter Michael

Cordillera del Paine, Chile

The Cordillera del Paine is part of Chile's Parque Nacional Torres del Paine in the southern Andean foothills. The white rock is the Cordillera del Paine granite, and the darker rocks at the summits and bases of peaks are sedimentary rocks of the Cero Toro formation.

A shallow sea covered this region about 100 million years ago. It deposited mud and sand that eventually became shale and conglomerates. Twelve million years ago, granitic magma intruded the sedimentary rocks from below. By studying the chemistry of the granite, we know that the magma solidified about 5 kilometres below Earth's surface. The dark rock capping the spires was the roof of that magma chamber, and dark rock at the base of the spires made up the walls of the chamber.

Since intrusion of the granite, the crust has been uplifted about 7 kilometres, and it has been deeply eroded in the recent past, mostly by glaciers that covered the area only a few thousand years ago. Today a vast ice sheet covers the Andes only 10 kilometres to the west; small glaciers below the granite spires are remnants of that once more extensive ice sheet. Rock pulverized by glaciers gives the surrounding lakes their light green, blue, and gray hues.

Photos by
Monte D. Wilson

Alpine Karst, Austria

In the northern Limestone Alps are several areas where chemical weathering (solution) of limestone has produced unusual surface features, caves, and underground drainage. This type of topography is called "karst," after the Karst Plateau of Yugoslavia, where it is especially well developed. The scenic Dachstein region near Hallstatt and Obertraun, Austria, is an alpine karst area. Rain and snowfall have caused extensive solution of the Triassic (210 to 230 million years old) limestones exposed over most of the Dachstein plateau. Water is diverted into underground drainage systems along intersecting fractures (joints) in the rock, and the influence of joints, faults, and bedding can be seen in caves. A deep valley formed by glaciers drains ground water from the northern part of the Dachstein plateau.

Photos by Thomas P. Miller,
courtesy of U.S.
Geological Survey

Aniakchak Crater, Alaska, U.S.A.

The volcanoes of southwestern Alaska form part of the Pacific "Ring of Fire"—volcanoes located over down-going lithospheric plates around the Pacific basin. One of the most spectacular volcanic landforms is Aniakchak Crater. The crater is a steep-walled, ice-free depression, 10 kilometres in diameter, and more than 1 kilometre deep. The basin-shaped depression, called a caldera, was formed about 3400 years ago by collapse of a large, deeply eroded and glaciated volcanic cone after a catastrophic eruption of more than 50 cubic kilometres of volcanic debris. Dense, hot flows of gas and ash from the eruption traveled more than 50 kilometres from the volcano rim. Airborne ash from this eruption was carried 1100 kilometres to the north. The volcano has been active since the caldera collapse; an eruption in 1931 deposited ash over a broad area of southwestern Alaska. Much of the vegetation inside the caldera was destroyed during this eruption and is only now making a comeback.

Photo by
Robert S. Anderson

Photo by
Robert S. Anderson

Earth Pillars, Utah, U.S.A.

Earth pillars are columns of rock debris, capped with large blocks of sandstone. The spectacular Straight Cliffs earth pillars in Utah rise up to 40 metres above the surrounding surface and sport capping blocks as much as 15 metres across. Here, the pillars form in the sides of gullies that are cut into major landslide deposits. The capping blocks appear to protect the material below them from water erosion.

Photo by
Robert S. Anderson

Photo by
Suzanne Prestrud Anderson

Photos by George H.
Denton, courtesy of
National Science
Foundation

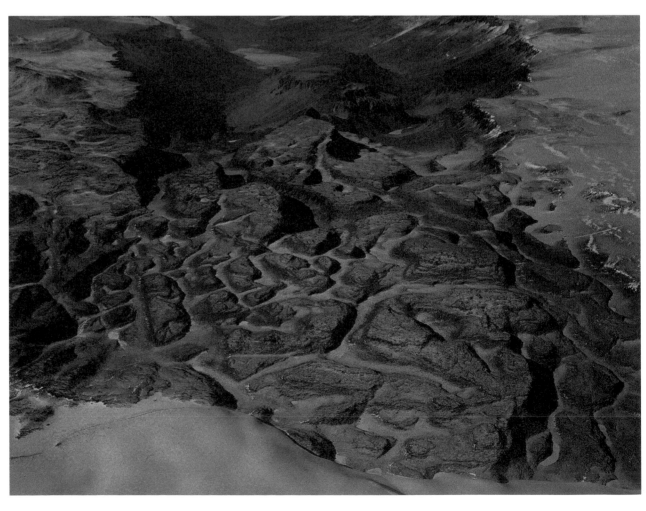

Glacial Land Forms, Transantarctic Mountains, Antarctica

T he Transantarctic Mountains, 2000 kilometres long, rise from the ice sheet covering most of the Antarctic continent. Over the past 10 million years or so, ice has occasionally covered the mountains, and stream channels have formed beneath the ice.

Mount St. Helens, Washington, U.S.A.

odern volcanoes of the Cascade Mountains of the northwestern United States and southwestern Canada are part of the Pacific "Ring of Fire" formed over down-going lithospheric slabs around the Pacific Ocean Basin. On May 18, 1980, a 5.1 magnitude earthquake shook Mount St. Helens in Washington State and triggered a major eruption. Four hundred metres of the volcano collapsed or were blown outward, leaving behind a 2550-metre-high beheaded peak and creating a horseshoe-shaped crater, 1.5 kilometres wide. Since the eruption, a new dome, 280 metres high, has grown inside the crater.

Photo by Lyn Topinka,
courtesy of U.S.
Geological Survey

Photo by Lyn Topinka,
courtesy of U.S.
Geological Survey

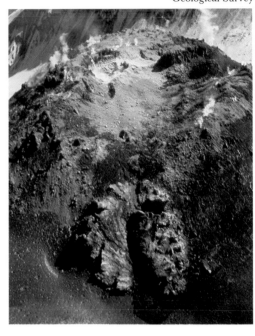

Trees in forests near volcanoes such as Mount St. Helens contain a wealth of information about the timing of volcanic eruptions. Major eruptions are recorded as abrupt thinning of annual rings in old trees in areas of heavy ash fall. It is also possible to date volcanic deposits by matching the ring-width variation patterns of dead trees within such deposits with those of living trees. These studies help determine how often a volcano has erupted in the past and perhaps how often it will erupt in the future.

Photo by
David K. Yamaguchi

Photo by Lyn Topinka,
courtesy of U.S.
Geological Survey

*Mount St. Helens,
Washington, U.S.A.*

Photo by Lyn Topinka,
courtesy of U.S.
Geological Survey

Photo by
David K. Yamaguchi

Folds on Venus

A radar image, taken from Arecibo, Puerto Rico, of Freyja Montes, a belt of deformed crust at the northern margin of Lakshmi Planum, on Venus. Radar-reflective bands, the thinnest of which are 15 to 20 kilometres wide, are interpreted to be folds. The folds are concentrated in a mountain belt that rises 3 to 4 kilometres above the surrounding plains, suggesting intense deformation similar to that in fold belts on Earth.

Photo by L. S. Crumpler, James W. Head, and Donald B. Campbell, courtesy of National Astronomy and Ionosphere Center, NSF/NASA

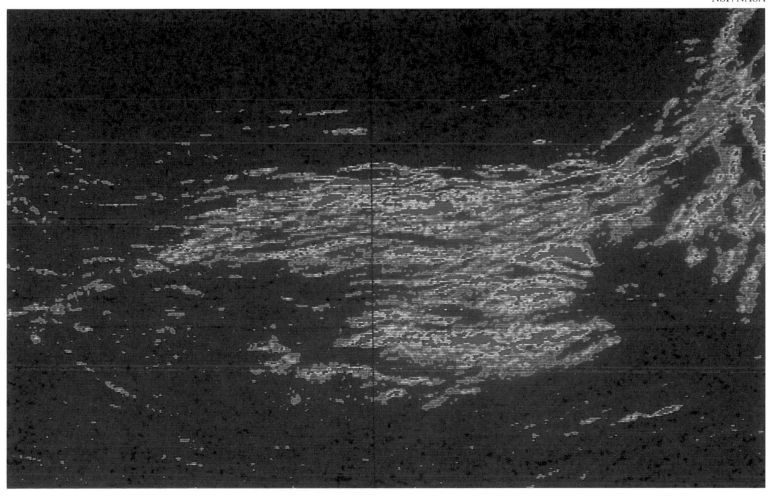

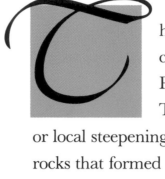

The Waterpocket Fold forms the core of Capitol Reef National Park in south-central Utah. The fold is a large monocline, or local steepening of otherwise gently dipping rocks that formed at the time of uplift of the Rocky Mountains, about 70 million years ago. The fold consists of steeply dipping sedimentary rocks that range in age from Permian to Cretaceous (late Paleozoic to late Mesozoic, about 260 to 70 million years old). These sedimentary rocks were deposited by water in marine basins or in continental streams and lakes, and by desert winds as giant sand dunes.

Erosion by wind and water has removed softer layers of rocks to expose the harder rocks of the fold. Slow but constant removal of grains over millions of years has sculpted the rocks of Capitol Reef into twisting canyons, rounded domes, and small natural depressions, or waterpockets; hence the name of the fold.

Photos by Russell F. Dubiel,
U.S. Geological Survey

Coastal Land Forms, Bahamas

The Bahamian Archipelago is a series of underwater platforms topped with more than 700 low islands, located in the tropical western North Atlantic Ocean. The archipelago extends 1400 kilometres from Little Bahama Bank off the coast of Florida, U.S.A., south to include the Turks and Caicos islands, near Hispaniola. The rocks and sediments of the archipelago consist of several forms of calcium carbonate. The islands are capped by limestones, and sea cliffs and inland ridges are composed of carbonate eolianite, windblown sand that accumulated as dunes and was later turned into rock. Eolianite is susceptible to solution by rain and by ground water and to erosion by wave action; the result is varied land forms, including sea arches and caves. The pinkish tint of the beaches comes from fragments of shells of a microscopic animal, *Homotrema rubra.*

Photos by
H. Allen Curran

San Andreas Fault, Carrizo Plain, California, U.S.A.

The San Andreas fault in California is part of the boundary between the Pacific and North American plates. In the arid Carrizo Plain, central California, several places show the interplay between gully erosion and land shifts along the fault. Movement on the fault has been intermittent, but over the past few million years it has averaged about 5 centimetres per year. As a result of these jerky displacements, older stream courses have been completely disrupted, while some younger ones jog to the right where they cross the fault trace. These offset streams show that movement along the fault has been essentially horizontal, that the far side has moved to the right, and that some episodes have involved at least 6 metres of displacement — enough to produce a major earthquake.

Photos by
John S. Shelton

Photos by
James McCalpin

Brooks Range, Alaska, U.S.A.

Glaciers in Pleistocene time (less than 1.6 million years ago) carved the U-shaped valleys of the central Brooks Range, Alaska. Glaciated peaks composed of Paleozoic (245 to 570 million years old) sedimentary rocks form a folded and faulted terrain, and Pleistocene valley glaciers gouged out basins, creating picturesque lakes such as Wild Lake at the head of the Wild River. Evidence of freezing and thawing and recent glacier movement can be seen along the Tinayguk River valley.

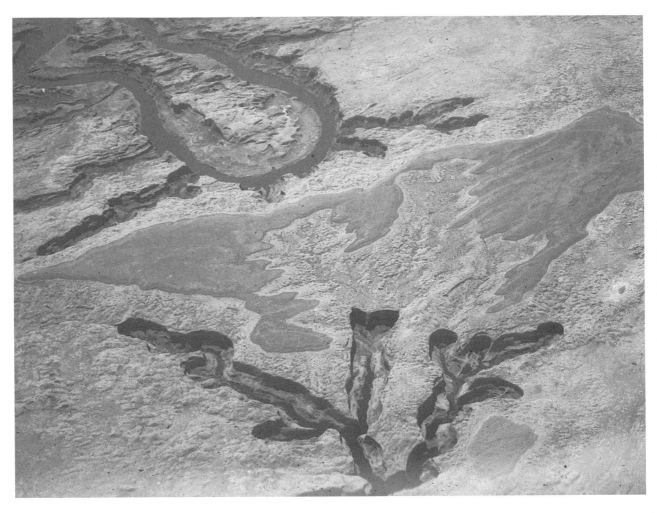

Photos by
Julie E. Laity

Ground-water Erosion, Utah, U.S.A.

The Glen Canyon region, Utah, contains many theater-headed valleys draining to the Colorado River and its tributaries. Many are in the Navajo Sandstone, an ancient dune-sand deposit. Ground-water erosion combined with surface flow helps to form the valleys. Ground water moves downward through the sandstone until it encounters the impermeable Kayenta Formation. Then it moves sideways through pores and fractures and emerges at seeps. This process may be aided in winter by freezing and thawing and the formation of icicles on seepage walls. Enlargement of cavities and alcoves undermines support for the cliff, and slabs fall off it at canyon-head walls and side walls.

Rodadero, Peru

he Rodadero is a hill with grooved and scratched (striated) wavy surfaces near ruins of the Incan fortress Sacsayhuamán, near Cuzco, Peru. Three theories have been advanced to explain the origin of this "toboggan slide": glacial, volcanic, and structural.

Formation by glacier seems unlikely. The nearest evidence of local glaciation occurs at least 250 metres above the Rodadero. Moreover, the grooves dip beneath limestone of Mesozoic age (65 to 245 million years old), much older than any nearby glaciation.

Some believe lava was extruded along an irregular fissure, and the scratches and grooves resulted from stretching of the lava as it oozed down the side of the hill. However, the hill is not volcanic rock.

Most likely, the grooves formed when blocks of rock moved past each other along a fault. The hill probably is the polished surface of an almost flat fault.

Photos by
Siegfried Muessig

Wasatch Folds, Utah, U.S.A.

Limestone of the Mississippian (about 320 to 345 million years old) Gardison and Deseret Formations in Rock Canyon, east of Provo, Utah, were folded about 80 to 100 million years ago. The limestone was exposed during the past 10 to 15 million years by rapid uplift of the Wasatch Mountains.

Shale has weathered more rapidly, leaving ridges of the more resistant limestone. Scrub oak and maple are concentrated in the less resistant zones, where soil and water are available. The maple trees (red and orange, upper center) are growing in one layer of rock, forming a line parallel to the fold.

Photos by
Bart Joseph Kowallis

Torfajökull Volcano, Iceland

Photos by
David W. McGarvie

The Torfajökull Mountains are in the uninhabited interior of southern Iceland. The area contains one of Iceland's largest and most intense high-temperature thermal fields, and is surrounded by young volcanic vents and lava flows. The thermal fields produce their own gentle microclimates, environments in which plants—mostly mosses—can grow. Such thermal areas are fertile oases within an otherwise barren landscape of lava, ash, and snow.

There have been 11 eruptions at Torfajökull during the past 10,000 years, the most recent in 1477 A.D. The lava is rich in silica and forms a black volcanic glass called obsidian. Large cavities can form when gas bubbles are trapped during cooling of the lava.

Because of its extensive thermal fields, Torfajökull is a prime site for the development and installation of geothermal power stations.

Erosional Forms, Northern Arizona, U.S.A.

I n the Painted Desert of northern Arizona, wind and water erosion of multicolored sandstone and shale of early Mesozoic age (about 200 million years old) has produced many unique and striking rock features.

Photos by
David A. Lawler

Aligned Coulees, Alberta, Canada

A ligned coulees in the valleys of the Oldman and Castle rivers, southern Alberta, Canada, may have resulted from a combination of strong southwesterly postglacial winds; subsurface joint and fault systems; the regional slope; trails made by prairie grazers, mainly bison, on their way to water; and differences in rock type.

Photos by
Chester B. Beaty

Photos by
Robert A. Baird

Folding and Cleavage, Appalachian Mountains, Virginia and Tennessee, U.S.A.

 **S**edimentary rocks are generally laid down in horizontal layers called bedding. If the beds are deposited within the stable inner part of the continent known as the craton, they can remain flat-lying for hundreds of millions of years. If they are deposited near the continental margin, where continental collision and mountain building takes place, they can be deformed into folds. Folds form when rock layers are buried to depths of 8 to 10 kilometres or more and are heated to 250°C or more. These high temperatures make the rocks soft and plastic so they bend, rather than break. When fine-grained sedimentary rocks such as mudstone or shale are folded, the clay and mica mineral grains in them align parallel to each other. When the rock is exposed after folding, it tends to break apart along these parallel alignments in a process called cleavage. Slate quarried for roofs and blackboards formed this way.

The Work of Hot Water, Western U.S.A.

When water flows from the earth surface down to a hot body of rock and back, hot springs, geysers, mud pots, and fumaroles result. As it passes through the rock the water dissolves material from it. When the water reaches the surface, it cools and deposits a special kind of limestone called travertine. Geysers are hot springs that store up enough energy to periodically erupt scalding water and steam. Terraces may grow several centimetres a year. Colors in the rocks and water come from bacteria, algae, and dissolved minerals.

Photo by
Niall J. Mateer

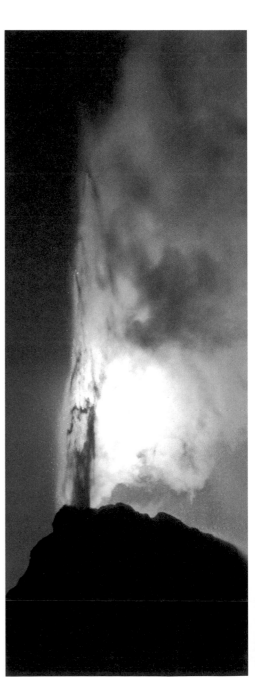

Photo by
Roderick A. Hutchinson

Photo by
Roderick A. Hutchinson

Photo by
Robin P. Diedrich

Photo by
Roderick A. Hutchinson

Photo by
Robin P. Diedrich

Photo by
Roderick A. Hutchinson

Photo by
Robin P. Diedrich

Photo by
Niall J. Mateer

Photo by
Roderick A. Hutchinson

Photo by
Roderick A. Hutchinson

Photo by
Niall J. Mateer

NASA Viking Orbiter
Digital Mosaic, courtesy
of U.S. Geological
Survey, Flagstaff,
Arizona

NASA Viking Orbiter
Digital Mosaic, courtesy
of U.S. Geological
Survey, Flagstaff,
Arizona

Valles Marineris, Mars

Mars is one of only two planets (the other is Venus) in our solar system that has many features similar to those on Earth: for example, areas that resemble deserts. This similarity is no coincidence: Mars is extremely dry, has a thin atmosphere, and is very cold (an average of $-60°C$). The Valles Marineris is a valley more than 2000 kilometres long and as deep as 7 kilometres. The valley has landslides, gullies, hills that were emplaced by floods, possible remains of lake beds or volcanic flows, and wind-sculpted features.

NASA Viking Orbiter
Digital Mosaic, courtesy
of U.S. Geological
Survey, Flagstaff,
Arizona

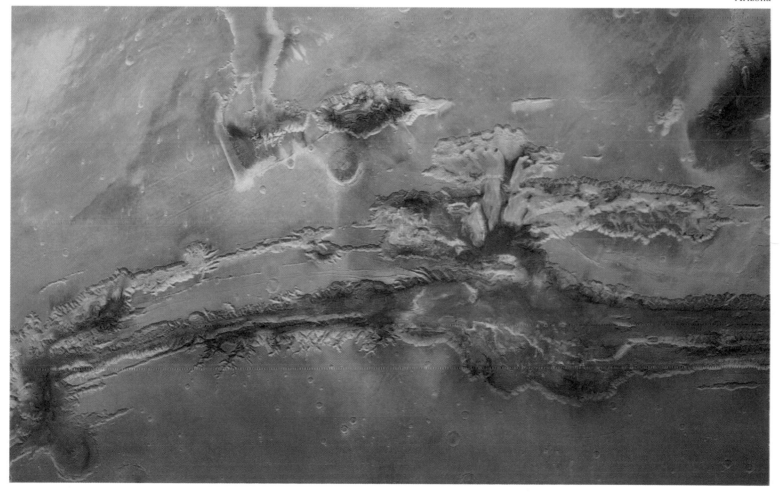

Tower Karst, China

**S**pectacular karst towers in southern China have inspired Chinese artists for centuries. The limestone that forms the pinnacles was deposited in Devonian to late Carboniferous (mid-late Paleozoic, 285 to 410 million years ago). Rainwater has dissolved some of the limestone, leaving the towers. Karst is named after the Krš Mountains of Yugoslavia.

Photo by
Manuel G. Bonilla

Photo by
Manuel G. Bonilla

Tower Karst, China

Photo by
Manuel G. Bonilla

Photo by
G. Shanmugam

Photo by
Manuel G. Bonilla

84

Rock from the Earth's Mantle

n places all over the world, molten rock (magma) has brought up solid fragments of Earth's mantle from depths as great as 150 kilometres. These rare solid fragments are found in kimberlite, the most common source of natural diamonds, and in other volcanic rock. Most fragments are peridotite, brought up and cooled so quickly that mineral compositions and textures have not changed. The compositions of the peridotite minerals indicate the depths and temperatures from which they came. The large grain–small grain textures indicate slow, solid flow of the rock at high temperature and pressure. This flow within Earth's interior results in the movement of tectonic plates on the surface. A polarizing microscope reveals, in various colors, the minerals in a thin slice of peridotite about 1 centimetre across from a diamond pipe in southern Africa.

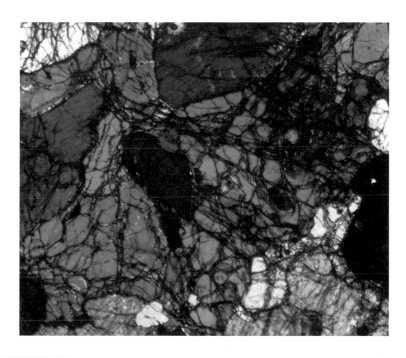

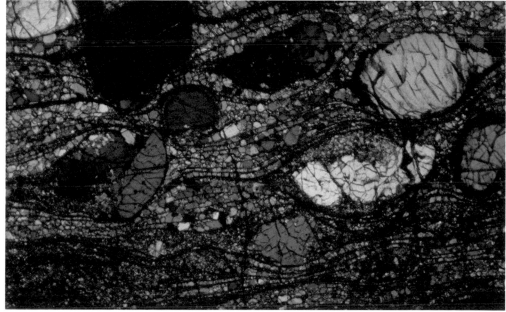

Photos by
Douglas Smith

Photos by
Martin J. Van Kranendonk

Salt, Arctic Canada

ircular and oval domes of salt shape the landscape of Axel Heiberg Island, Arctic Canada. These forms are composed primarily of rock salt with minor amounts of anhydrite and/or gypsum. This salt was deposited about 330 million years ago and formed thick beds over a vast area. They were later covered by marine sediment, the weight of which forced the older, more buoyant salt to rise. Over millions of years, the salt pushed through thousands of metres of sediment until it reached the surface. These rocks originally formed near the equator and drifted northward to their present location.

Photo by
David R. Dockstader

Photo by Grant C. Willis,
courtesy of Utah Geological
and Mineral Survey

Arches National Park, Utah, U.S.A.

Wind and water have sculpted sandstone rocks to form the arches of Arches National Park, Utah. Soft, almost fluid layers of salt underlie the more rigid sandstone, which formed from 200-million-year-old sand dunes bonded with red iron oxide cement. Movement in the salt caused parallel fractures (joints) to form in the sandstone. Erosion removed material along the joints, leaving sandstone walls or fins. Holes that formed where the cement was weak have enlarged to form the arches we see today.

Photo by
David R. Dockstader

Photo by Grant C. Willis,
courtesy of Utah Geological
and Mineral Survey

Photo by
David R. Dockstader

Photo by Grant C. Willis,
courtesy of Utah Geological
and Mineral Survey

Photo by
David R. Dockstader

90

Photo by Grant C. Willis,
courtesy of Utah Geological
and Mineral Survey

Arches National Park, Utah, U.S.A.

Photo by
David R. Dockstader

Photos by
Gail L. Marquardt

Coastal Features, California, U.S.A.

Point Lobos, on the Pacific shore south of Carmel, California, is an exposure of a rock unit, the Carmelo formation, which tells the story of a submerging and emerging coastline.

During submergence 60 million years ago, ocean currents and waves caused deposition of pebbles, rocks, and sand in the shallow coastal sea.

Today the coastline in the Point Lobos area is emerging or has recently emerged, exposing the ancient deposits. Coastal erosion has created interesting and odd-shaped formations on the beach.

Suez Rift, Egypt

The Gulf of Suez separates the African continent from the microcontinent of the Sinai Peninsula and offers an excellent example of the early stages of continental rifting. The Sinai Peninsula began to separate from Africa 30 million years ago. At that time Sinai was part of the Arabian continent and the Gulf of Suez was the northern extension of the widening Red Sea separating Arabia from Africa. Sinai split from Arabia along the Gulf of Aqaba–Dead Sea fault system 20 million years ago, and extension in the Gulf of Suez stalled. Sinai is no longer separating from Africa, and the Arabian peninsula is drifting away from both; oceanic crust is filling the void beneath the Red Sea. Formerly horizontal rock strata have been tilted by faulting along the margins of the gulf.

Photos by
Dana Q. Coffield

94

Popocatépetl Volcano, Mexico

opocatépetl, Mexico, is the "smoking mountain" of the Aztecs and the fifth highest mountain in North America. This volcano is part of the Pacific "Ring of Fire," located over subducting plates. It is snow clad year-round and has glaciers on its flanks. Although it last erupted in 1920, sulfur fumes still emanate from cracks in the crater. Francisco Mantano and Spanish soldiers seeking sulfur for gunpowder first climbed the mountain in 1523.

Photos by
Dana Q. Coffield

Central Park, New York City, U.S.A.

Glacially polished and scratched schist and gneiss of the Hartland Formation, Central Park, New York City, reveal a complex geologic history. Metamorphosed Cambrian-Ordovician (early Paleozoic, 450 to 570 million years old) sedimentary rocks were deformed as an oceanic island chain collided with the margin of North America 450 million years ago. Well-layered rock outcrops display complex folds intruded by granitic dikes. The effects of glaciation (10,000 to 25,000 years ago) include boulders brought from far away, as well as deep southeast-trending furrows and scratches carved into the rock. The skyline of New York City reflects its geology.

In midtown Manhattan and around Wall Street, bedrock is at or near the surface and can support the weight of the sky-scrapers which are built there.

Photos by
Bruce Taterka

Mining Diamonds, Brazil

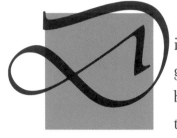 iamonds are among the most highly valued gemstones. They are usually colorless; yellow, blue, red, or brown tints are caused by impurities. They are the hardest substance known on Earth: blackish, poor-quality stones and diamond chips are used in drill bits, saw blades, and other cutting devices.

Diamonds are formed under high pressure and great heat. They are primarily found in kimberlite intrusions—plugs of once molten rock from deep within the earth. In Brazil, kimberlite plugs either have been eroded away or are covered by younger layers and are mined from river sediments or Precambrian (older than 570 million years) conglomerates. Because of the extreme hardness and inert chemical composition of diamonds, they are only slightly affected by the abrasive forces of wind and water. Thus they can travel for hundreds of miles with stream sediments or ocean currents without any significant change.

Photos by Winsried Schmidt and A. Hoppe

Copper, Arizona, U.S.A.

Photo by Sterling S. Cook,
courtesy of U.S.
Bureau of Mines

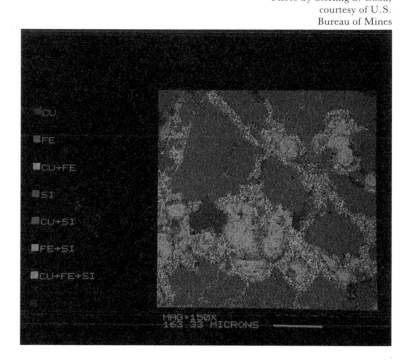

opper deposits in the western United States formed in granite that rose into Earth's upper crust. The Silver Bell copper mining district is northwest of Tucson, Arizona.

An X-ray "map" of copper ore from the Santa Cruz deposit in Arizona shows the mineral atacamite (red) filling cracks and pores between grains of quartz (blue) and limonite (green).

Photo by
David A. Sawyer

Great Sand Dunes, Colorado, U.S.A.

The Great Sand Dunes of south-central Colorado are among the tallest dunes in the world: more than 200 metres from base to top. The wind-deposited dunes are located in the temperate, arid San Luis Valley.

As winds move from the southwest across the valley, they pick up sand and silt and carry it toward the neighboring Sangre de Cristo Mountains. The mountains slow the winds, which drop the sand and silt. Occasional storms bring winds from the northeast and east and stack the sand back onto itself, giving the Great Sand Dunes their tremendous height.

Photos by
Peter K. Blomquist

2

1

Glaciers

s warm moist air rises over mountain ranges and cools, rain forms at lower elevations and snow at higher elevations. As the snow accumulates, it slowly changes to ice. Pressure builds in the accumulating ice and causes it to flow, moving down valleys as frozen rivers, or glaciers, at a rate of centimetres to metres per day.

The ice sculpts the mountains, forming U-shaped valleys; narrow, sawtooth ridges; and spirelike peaks. In the process, glaciers pick up rock debris, which rides on the edges of the glacier. Deposits of this debris are called moraines.

Examples of Glacial Features
1. Rongbuk glacier, with Mount Everest in the background. 2. Avalanche above Baltoro glacier, Karakoram Mountains, Pakistan. 3. Cirque below Pukajirka, Cordillera Blanca, Peru.
4. Moraines and crevasses on the Mer de Glace, Mont Blanc, France.

Photos by
Dana Q. Coffield

3

4

Photo by Rodney
Catanach, Kenneth A.
Rasmussen, and Carlton
E. Brett

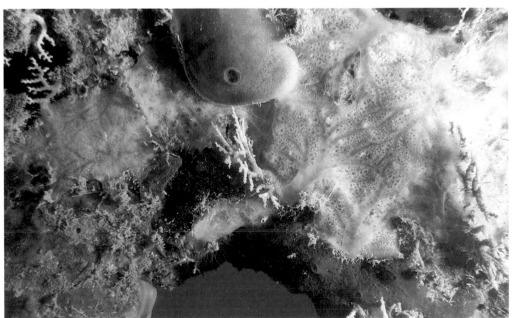

Photo by Ivan P. Gill,
Kenneth A. Rasmussen,
and Carlton E. Brett

Modern Reef Community, Virgin Islands, U.S.A.

Photo by Kenneth A. Rasmussen and Carlton E. Brett

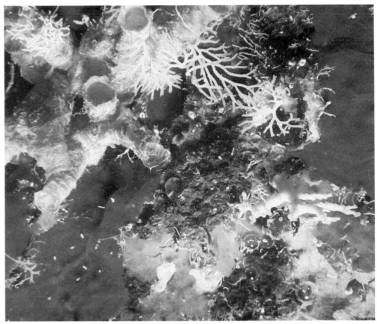

heltered areas in the marine realm have long provided a habitat for shade-loving organisms that live beneath corals, shells, and rocks. These organisms form seldom-seen, beautiful, diverse communities. Surveys of reef caves and coral plates in Salt River Submarine Canyon, St. Croix, U.S. Virgin Islands, provide new insights into which organisms are likely to be preserved as fossils and which are not.

Photo by Kenneth A. Rasmussen and Carlton E. Brett

Bighorn Dolomite, Wyoming, U.S.A.

The 450-million-year-old Bighorn Dolomite forms spectacular escarpments in the Rocky Mountains of northwest Wyoming, such as at Steamboat Point in the Bighorn Mountains.

The dolomite contains an ancient animal-burrowing network called *Thalassinoides*. This kind of burrowing and associated fossils, such as gastropods, brachiopods, corals, and crinoids, indicate that the original calcium carbonate sediments were deposited in a marine environment.

Note the characteristic rough-weathering surfaces.

Photos by
Donald H. Zenger

Photo by
Gerhard Wörner

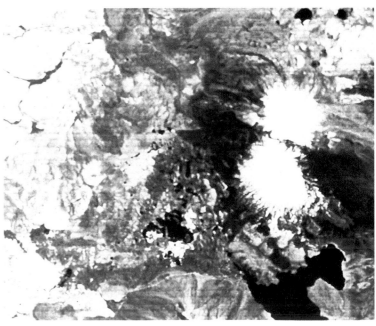

Photo by LANDSAT
courtesy of P. Francis

Volcan Parinacota, Chile

Twin volcanoes Parinacota and Pomerape in northern Chile are yet another part of the Pacific "Ring of Fire." The continental crust here is about 70 kilometres thick, twice the thickness of typical crust, and so the volcanoes rise from the Altiplano of northern Chile at an elevation of about 4500 metres to summits that are over 6000 metres. The satellite view shows the volcanoes in white, vegetation in red, and lakes in black.

About 13,500 years ago, the cone of Parinacota collapsed, causing a debris avalanche similar to that of the 1980 Mount St. Helens eruption. Lava flows have rebuilt the present snow-covered cone of the volcano.

Photo by
Jon P. Davidson

Photo by
Nancy J. McMillan

Photos by
Ronald C. Blakey

Southern Ramparts, Colorado Plateau, Arizona, U.S.A.

The Mogollon Rim is a nearly continuous escarpment along the southern edge of the Colorado Plateau from central to eastern Arizona. About 250 million years ago, wind and ocean tide deposited silt and sand, now red and white siltstone and sandstone. The southern Colorado Plateau rose several times, most recently about 5 to 15 million years ago. Streams have cut into the plateau to form the remarkable landscape.

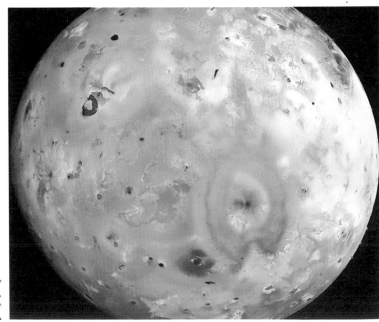

Photos submitted by
Alfred S. McEwen,
courtesy U.S. Geological Survey
and NASA

Pele Eruption, Io

O n March of 1979, *Voyager 1* flew past Jupiter and returned closeup pictures of the moons orbiting the giant planet. Active volcanism on Jupiter's moon Io, driven by tidal interactions with Jupiter and the other large satellites, was one of the most exciting discoveries of the Voyager missions. The volcanic activity was first detected from an image showing a 300-kilometre-high plume above Io's surface. This plume, named Pele after the Hawaiian goddess of fire, was the largest of nine active plumes detected by *Voyager 1*. When the *Voyager 2* spacecraft arrived at Jupiter four months later, Pele was no longer active.

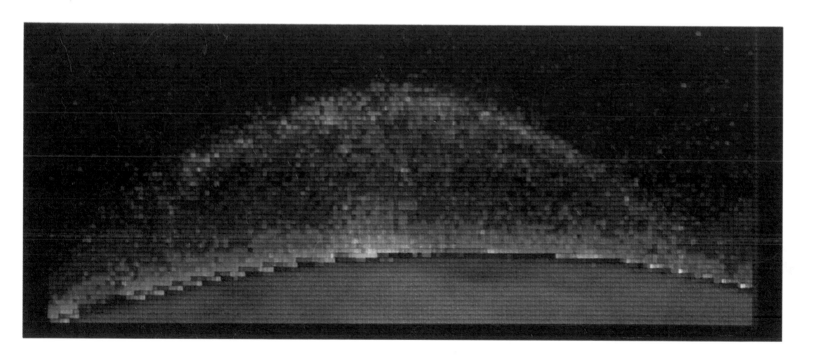

Colca Canyon, Peru

 olca Canyon of southern Peru, deepest canyon in the world, is twice as deep as the Grand Canyon of the United States. The Colca River cuts through a thick section of Late Jurassic and Early Cretaceous age (late Mesozoic, approximately 100 to 160 million years old) rock that originally accumulated as continuous, flat-lying layers of sediment. The sedimentary rocks were deformed into spectacular folds.

This area was uplifted and eroded after the deformation that caused the folding. Throughout the canyon young deposits overlie the folded rocks. The river has been downcutting at the rapid rate of about 2 millimetres each year over the past 250,000 years.

Photos by
William J. Devlin

Photos by
Dennis K. Bird,
Craig E. Manning, and
Nicholas M. Rose

Skaergaard Complex, Greenland—
A Product of Continental Rifting

ifting of Greenland from
Eurasia began about 60 million
years ago with the eruption of a
thick sequence of basaltic lava,
followed by a long period of dike emplacement.
These dikes represent the formation of new crust
at the continental margin, and the eroded cores of
magma chambers within and near the dike complex are the deep roots of ancient volcanoes. The
Skaergaard intrusion, shown here, is an example
of rocks formed in one of these magma chambers.
Ground waters migrated through cracks in the
cooling intrusive rocks and fed geysers and hot
springs at the surface. The geologic processes that
occurred in East Greenland 50 to 60 million years
ago can be observed today in the volcanoes, hot
springs, and geysers of Iceland.

Ductile Deformation, Wyoming, U.S.A.

eformed gneiss, about 2.5 to 2.9 billion years old, of the Teton Range, Wyoming, shows different kinds of folds. These kinds of rock folding can happen only at high temperatures, when the rocks can flow easily, at great depths within Earth's crust. Later uplift and erosion has exposed them at the surface.

Photos by
Scott H. Miller

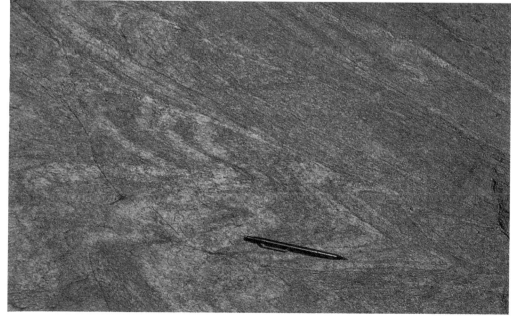

Photos by
William J. Devlin

Okmok Caldera, Alaska, U.S.A.

kmok Volcano, on Umnak Island, in the Aleutian Islands, has gentle slopes and is symmetrical and circular. A 10-kilometre-wide, 500-metre-deep, water-filled crater now caps the volcano.

Okmok Volcano is still active today. Steam comes out of cones within the crater. A volcanic eruption sometime after 1959 produced the black lava flow.

Photo by Martin G. Miller
and Mark Savoca

Death Valley, California, U.S.A.

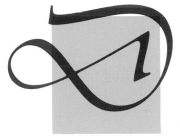eath Valley continues to form as western North America pulls away from the rest of the continent. Ancient and modern borax-bearing deposits have been mined since 1881. The floor of Death Valley is below sea level. Badwater, in the central valley, is the lowest point in the United States.

Photo by
D. W. Caldwell

Photo by
Peter K. Blomquist

Red Rocks Park, Colorado, U.S.A.

Red Rocks Park, near Denver, Colorado, has rocks ranging in age from Precambrian to Cenozoic (1700 to 1.6 million years old). The Fountain Formation forms the picturesque flatirons. The red sandstone, silt-stone, claystone, and conglomerates were deposited 300 million years ago by desert streams that flowed from an ancient mountain range, the Ancestral Rocky Mountains.

Photos by
Manuel G. Bonilla

Iguazu Falls, Argentina-Brazil

Most of the water of Iguazu Falls, on the border between Argentina and Brazil, is concentrated along a narrow band called Devil's Throat. Erosion there has caused the falls to retreat upstream some 20 kilometres from the Iguazu River's confluence with the much larger Paraná River.

The rocks at the falls are 200-million-year-old lava flows dating from the beginning of rifting of South America away from Africa.

Inclusions in Diamonds

Many diamonds contain gemlike mineral inclusions that help us understand how diamonds formed. These minute crystals provide valuable information on distribution of chemical elements in Earth's interior, where diamonds crystallized. Some common inclusions are colorless olivine, reddish-brown chromite, ruby-red chrome magnesian garnet, emerald-green chrome clinopyroxene, yellowish-brown magnesium-iron-calcium garnet, and black graphite.

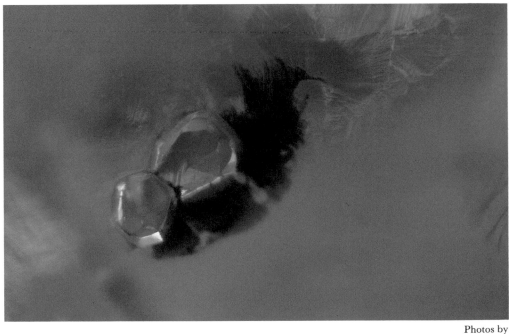

Photos by
Irene S. Leung

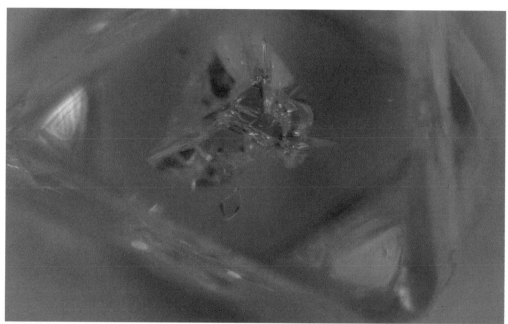

Saddlebag Lake Pendant, Sierra Nevada, California, U.S.A.

The Saddlebag Lake Pendant (a remnant of rock that formed before the intrusion of granitic rocks) is at the east edge of Yosemite National Park, California. The dark rocks of the pendant are about 100 to 500 million years old and consist of metamorphosed sedimentary and volcanic rocks. The oldest rocks were attached to North America about 360 million years ago. The lighter colored rocks are 80- to 90-million-year-old granites which formed the core of the Sierra Nevada.

Lake Canyon is a classic U-shaped valley carved by glaciers. It is actually a hanging valley that ends 300 metres above the main canyon.

Basaltic pillow lava formed during submarine volcanic eruptions.

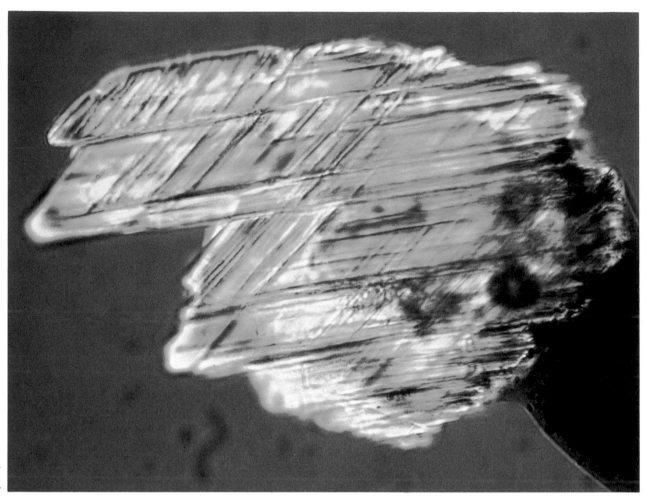

Shock-Metamorphosed Quartz

he rare metal iridium is unusually abundant in marine sedimentary rocks that span the Cretaceous-Tertiary (K-T; Mesozoic-Cenozoic) boundary (65 million years old) in Italy, Denmark, and New Zealand. This boundary coincides with the most extensive extinction of animal life known in the geologic record. Some workers think that an enormous asteroid, about 10 kilometres in diameter, struck Earth, causing the iridium-rich deposits and extinction of many life groups, including dinosaurs.

Some quartz and feldspar mineral grains in K-T boundary sediments support this idea. The grains contain many closely spaced, more or less parallel fractures formed by high-pressure shock waves. In a microscopic view of a thin, transparent slice of one mineral, the microscopic fractures appear as dark lines.

The abundance and large size of these "shocked" quartz and feldspar grains from North America suggest that if an asteroid impact did occur, it was probably close to that continent.

Deep Marine Deposits, Antarctic Peninsula

During Mesozoic time (65 to 225 million years ago), what is now the Antarctic Peninsula was the site of an active chain of volcanoes (an island arc). Rocks from this island arc were deposited in rivers, deltas, and the deep ocean.

In these deposits, white sandstone and black siltstone alternate, producing so-called zebra rock, which is later deformed by slumping and faulting.

Photos by
Peter J. Butterworth

133

Photos by
Dana Q. Coffield

Continental Collision

 ll of the world's mountains over 8000 metres high are located in the vast range of the Himalayas. Over the past 50 million years, the Indian subcontinent has slowly, at a rate measured in centimetres per year, collided with the Asian continent; the Himalayas in the collision zone are being driven to ever greater elevations.

The void between India and Asia was once occupied by an ocean. As India drifted north toward Asia, the oceanic crust disappeared beneath Asia.

The Karakoram Range of Pakistan shows the effects of this continental collision. It includes K-2, the second (or first?) highest mountain on Earth, and the Trango Towers.

Closeups show folded metamorphosed sediments and an example of rock layers that have been pulled apart, the spaces between them filled by white calcite.

Photos by
Roderick A. Hutchinson

Valley of Ten Thousand Smokes, Katmai, Alaska, U.S.A.

On June 6, 1912, the eruption of Novarupta Volcano, ten times greater than the 1980 eruption of Mount St. Helens, devastated the Katmai region in southwest Alaska. Large volumes of hot, incandescent ash and pumice were ejected from Novarupta and filled the 65 square kilometres of the upper Ukak River valley, destroying all life.

Innumerable vents and fissures developed in the hot volcanic ash deposits, allowing steam and gases to escape; explorers named it the Valley of Ten Thousand Smokes. Since the volcanic eruption, the thousands of fumaroles and gas vents have disappeared. Yellow, red, brown, black, and white patterns in the volcanic rock indicate areas where minerals have been deposited or have been removed by hot water. Deep, narrow canyons are being cut in the volcanic rocks.

Lost River Range, Idaho, U.S.A.

he Lost River Range in Idaho is a mountain block uplifted by faulting. There has been recent major seismic activity along the range-front fault: in 1983, a magnitude 7.3 earthquake shook the central part of the range.

The rocks are folded Devonian to Pennsylvanian (mid- to late Paleozoic, 290 to 360 million years old) strata of the central part of the Lost River Range. They were deformed about 80 to 100 million years ago.

Photos by
Thomas M. Jacob

Devils Tower, Wyoming, U.S.A.

evils Tower rises almost 300 metres above the low, rolling hills of northeastern Wyoming. Columns such as Devils Tower typically form when hot, liquid rock cools uniformly and moderately slowly; as the liquid loses heat to the colder enclosing rocks, it solidifies, shrinks, and cracks into many-sided columns perpendicular to the cooling surfaces.

Photo by
Wallace D. Kleck